Looking at Clouds

Fluffy Cumulus Clouds

By Kim Thompson

TABLE OF CONTENTS

A Crabtree Seedlings Book

Crabtree Publishing

crabtreebooks.com

School-to-Home Support for Caregivers and Teachers

This book helps children grow by letting them practice reading. Here are a few guiding questions to help the reader build comprehension skills. Possible answers appear here in red.

Before Reading:

- What do I know about clouds?
    - *I know that the sky has different clouds at different times.*
    - *I know that some clouds are storm clouds.*
- What do I want to learn about clouds?
    - *I want to learn about different kinds of clouds.*
    - *I want to know what the weather will be like by looking at clouds.*

During Reading:

- What have I learned so far?
    - *I have learned that clouds are made of water droplets or ice crystals.*
    - *I have learned that cumulus clouds look like cotton balls.*
- I wonder why...
    - *I wonder why hail sometimes falls from clouds.*
    - *I wonder why some clouds are gray.*

After Reading:

- What did I learn about cumulus clouds?
    - *I have learned that the weather is usually sunny when you see cumulus clouds.*
    - *I have learned that cumulus clouds are always changing.*
- Read the book again and look for the glossary words.
    - *I see the word **hail** on page 15 and the word **drizzle** on page 16. The other vocabulary words are found on page 22.*

Fluffy Cumulus Clouds

Clouds are made of tiny droplets of water or ice. One kind of cloud is called cumulus.

Cumulus clouds are bright white. They have fluffy tops and flat bottoms. They look like floating cotton balls!

Cloud Watch

If you see cumulus clouds, the weather is usually nice and sunny.

Water heated by the Sun turns into an invisible gas called **water vapor**. The gas rises up high. It cools and turns back into water droplets. They join together to make clouds.

Cloud Watch

Clouds are part of Earth's water cycle. Water moves from oceans, up to the sky to form clouds, then back down to the ground and the oceans.

Cumulus clouds form at a height where the air is warm enough that the water droplets stay liquid. Cumulus clouds grow as they collect more water.

As the cloud's water droplets grow bigger, they get too heavy to float. They fall from the cloud as rain or snow. This is called **precipitation**.

Cumulus clouds are always changing. The wind can push them lower or blow them higher.

Cumulus clouds can have different names.

Thunderstorms come from cumulonimbus clouds. These are cumulus clouds that grow tall and dark. They are full of water and bring rain, lightning, and thunder.

Cloud Watch

*Cumulonimbus clouds are the only clouds that can make **hail**.*

Stratocumulus clouds are low in the sky. They are lumpy and gray, with blue sky between them. These clouds bring **drizzle**. Stratocumulus clouds can be seen before and after storms.

Cloud Watch

Stratocumulus clouds can form in rows.

Altocumulus clouds are midway up in the sky. Rain may be on the way if you see these clouds.

Cloud Watch

Altocumulus clouds look like a flock of light-gray sheep!

Cirrocumulus clouds are high in the sky. The air is very cold. The water droplets freeze into ice **crystals**. Cirrocumulus clouds are a sign of sunny weather.

Cloud Watch

Cirrocumulus clouds can look like fish scales. A sky full of these clouds is called a ***mackerel*** *sky.*

Glossary

crystals (KRIS-tls): Small pieces of ice that float in the air in cold weather

drizzle (DRIZ-uhl): Light, misty rain

hail (hayl): Balls of ice that fall from the sky

mackerel (MAK-er-uhl): A fish often caught for food

precipitation (pri-sip-i-TAY-shuhn): The falling of water from the sky in the form of rain, sleet, hail, or snow

water vapor (WAW-tur VAY-pur): Water that has changed from a liquid into a gas

Pronunciation Guide

cumulus (KYOO-myuh-luhs)
cumulonimbus (KYOO-myuh-loh-NIM-buhs)
stratocumulus (stray-toh-KYOO-myuh-luhs)
altocumulus (al-toh-KYOO-myuh-luhs)
cirrocumulus (sir-oh-KYOO-myuh-luhs)

Cloud levels		Cloud types
High clouds *Cloud names that begin with "cirro" are the highest clouds in the sky.*	*above 18,000 feet (5,486 meters)*	**High clouds** *cirrus* *cirrostratus* *cirrocumulus*
Middle clouds *Clouds in the middle of the sky have names that start with "alto."*	*6,500 to 18,000 feet (1,981 to 5,486 meters)*	**Middle clouds** *altostratus* *altocumulus*
Low clouds *Clouds that form low in the sky have "strato" in their name.*	*up to 6,500 feet (1,981 meters)*	**Low clouds** *stratus* *stratocumulus* *nimbostratus*

Index

Written by: Kim Thompson
Designed by: Samara Parent
Series Development: James Earley
Proofreader: Kathy Middleton
Educational Consultant: Marie Lemke M.Ed.
Photographs: All images from Shutterstock

Crabtree Publishing

crabtreebooks.com 800-387-7650

Printed in the USA/CP2025

Published in Canada
Crabtree Publishing
616 Welland Avenue
St. Catharines, Ontario
L2M 5V6

Published in the United States
Crabtree Publishing
347 Fifth Avenue
Suite 1402-145
New York, New York, 10016

Library and Archives Canada Cataloguing in Publication
Available at Library and Archives Canada

Library of Congress Cataloging-in-Publication Data
Available at the Library of Congress

Hardcover: 978-1-0398-8123-5
Paperback: 978-1-0398-8484-7
Ebook (pdf): 978-1-0398-8243-0
Epub: 978-1-0398-8363-5
Read-Along: 978-1-0398-8568-4
Audio: 978-1-0398-8526-4